EUROPEAN ROUND TABLE ON INDUSTRY, SOCIAL DIALOGUE AND SUSTAINABILITY

A General Report

EF/93/10/EN

 European Foundation
for the Improvement of
Living and Working Conditions

EUROPEAN ROUND TABLE ON INDUSTRY, SOCIAL DIALOGUE AND SUSTAINABILITY

A General Report

by
ECOTEC Research and Consulting Limited

Loughlinstown House, Shankill, Co. Dublin, Ireland
Tel: +353 1 282 68 88 Fax: +353 1 282 64 56 Telex: 30726 EURF EI

Cataloguing data can be found at the end of this publication

Luxembourg: Office for Official Publications of the European Communities, 1993

ISBN 92-826-5817-1

© European Foundation for the Improvement of Living and Working Conditions, 1993.

For rights of translation or reproduction, applications should be made to the Director, European Foundation for the Improvement of Living and Working Conditions, Loughlinstown House, Shankill, Co. Dublin, Ireland.

Printed in Ireland

PREFACE

This was the third round table at which the social partners discussed questions related to the environment. The first one, held in Dublin in June 1988, was concerned with **the Role of the Social Partners in Improving the Environment** and concluded with the identification of key environmental issues in which the social partners are playing, or could play, an important role. Some of the recommendations of this round table were combined in the study on **Education and Training of Categories of Personnel concerned with Environmental Issues relating to industry,** carried out in 1989-90. A cooperation with CEDEFOP was established in relation to its project on occupational and qualification structures in the field of environmental protection.

In 1990, the European Foundation, the Hans-Böckler-Foundation and the Friedrich-Ebert-Foundation started a project on **Industrial Relations and the Environment in the European Community**, to explore the situation in the Community on this topic. The following questions have been addressed: How does management react within the company to environmental issues? How are unions and workers' representatives being informed; do they have a policy? Are the issues discussed between the two sides of industry? What issues are discussed (pollution, products, health and safety, security and employment)? Do these discussions take place within the established institutions of industrial relations or are new institutions emerging? Are the health and safety committees dealing with the external environment relating to the firms? A round table on this subject took place in Brussels, in November 1991. An important result of this first event was an improved understanding between UNICE and ETUC regarding the issues on which there is a common position and those on which there are important differences. The final results of this project based on a comparison of 10 countries and an analysis of 7 case studies were presented at the round table in Rome.

In 1991, the Foundation initiated the project on **The Firm and its Local Environment** which focuses on the ways in which companies, notably SMEs, and the social partners can become more involved in environmental protection and improvements and cooperate with local communities and relevant bodies. The first result of this project was also presented and discussed in Rome.

Eco auditing is an innovative approach in industry and the European Commission is promoting projects in this field. The Commission's efforts in eco-auditing were presented in Rome. **The Post-Rio reality and concerns** for industry and social dialogue were discussed in two panels. The first was composed of representatives of the United Nations, OECD, ILO and the International Association of Chambers of Commerce. The European aspects of the same subject were discussed by representatives of the European Commission, Governments, employers' associations and trade unions.

We believe that the general report we now publish reflects the efforts undertaken by all sides and their willingness to achieve a sustainable environment.

Clive Purkiss
Director

Eric Verborgh
Deputy Director

CONTENTS

Page

EXECUTIVE SUMMARY

MAIN PROCEEDINGS OF THE ROUND TABLE 1

Theme A : The Firm in its Local Environment 3

Theme B : Education and Training 10

Theme C : Industrial Relations and the Environment 14

Theme D : Eco-Auditing 19

Theme E : Panel Discussion 24

ANNEX A : European Round Table on "Industry, Social Dialogue and Sustainability" : Programme and Participants

ANNEX B : A Background Paper for the Round Table on Industry, Social Dialogue and Sustainability

ROUND TABLE ON INDUSTRY, SOCIAL DIALOGUE AND SUSTAINABILITY

EXECUTIVE SUMMARY

I. THE FOCUS OF THE ROUND TABLE

The Round Table on Industry, Social Dialogue and Sustainability, held in Rome was the third in a series of Round Tables which have sought to clarify and diffuse an understanding of the fundamental importance of social dialogue to achieving real improvements in living and working conditions in Europe.

The Rome Round Table built upon the work and conclusion of the previous Round Tables, as well as the continuing research programme of the European Foundation, to further develop social dialogue by focusing more specifically upon the **means** of achieving sustainable development, rather than focusing on defining broad statements of objectives.

Thus the Round Table sought to demonstrate the importance of social dialogue in contributing to:

- an awareness of environmental issues and their significance for industrial activity, amongst managers, workers, government and the public;

- a more informed response by industry to environmental pressures;

- a more efficient environmental performance by industry through improving the external relationships with different interest groups (customers, suppliers, regulations) necessary to conduct business, and improving internal industrial relationships necessary to secure the appropriate corporate response.

The Round Table was invited to focus, therefore, on:

- the role and influence of local co-operation initiatives as a vehicle for social dialogue;

- the need for additional initiatives and techniques for resolving conflicts and establishing consensus;

- the necessary management and procedural changes which are required by firms to respond effectively to environmental pressures;

- the necessary changes in industrial relations;

- the timescales which should be used when considering changes and the development of new policies;

- the issues and opportunities which need to be addressed in securing improvements of the environmental performance of industry in Objective One regions.

The programme and participants of the Round Table are presented in Annex A. As an initial contribution to the Round Table, some of the key issues which underlie the debate, together with some observations on possible policy directions were presented in a Background Paper. This Paper is reproduced as Annex B.

II. CONCLUSIONS OF THE ROUND TABLE

The first and fundamental conclusion to result was unanimity amongst participants that change towards a more environmentally benign system of production and social relations was both desirable and necessary and that social dialogue was fundamental to achieving this change. This conclusion was essentially the starting point for the discussion and was followed by success in identifying the current position of the policy debate in relation to the attainment of sustainable development, as perceived by a wide range of governmental and non-governmental organisations, social partners and other stakeholders (eg. consumers, local communities).

Despite the variety of interests and differing experiences of the issues involved there was agreement on a number of central questions, which can serve to provide a fresh and progressive platform from which to continue the development of policies and actions directed to securing sustainable development.

II.I MULTI-DIMENSIONAL CONTEXT

The Round Table considered various definitions of *"industry"*, *"social dialogue"* and *"sustainability"* and recognised that there were many different interpretations reflecting the wide range of stakeholders and their differing constituency of interests. It was not considered necessary nor practical to have single definitions of these terms. The complexity of interactions between the stakeholders and different spatial, sectoral and corporate levels must be recognised and that as a consequence no single "model" for securing the required change can be defined since change needs to take place at all levels. What emerged as central was the interdependency of change at different levels (eg. between securing improvements in the environmental performance of smaller local manufacturing firms and the development of supra-national frameworks in relation to say economic instruments), and the current lack of definition of this interdependency.

The realisation and acceptance of this multi-dimensional context was viewed as both part of the problem of attaining sustainability and part of the solution. The complexity and vast range of

Round Table on Industry, Social Dialogue and Sustainability

interests and perceptions, is clearly an obstacle to speedy and radical changes in social and production relations. However, recognising the complexity enables a more realistic assessment of the timescales and the processes necessary to achieve change. This assessment reduces the frustration at an apparent lack of progress in defining actions and securing the necessary change, which was felt by some participants, and opens up a large number of different routes for exploration and for making progress.

A key feature of the Round Table was the extension of the definition of interests in sustainable development from those of government and the social partners (industry employers, managers, workers) to include other stakeholders. Explicit mention was made of the need to define and integrate the interests of other stakeholders (eg. the public, consumers, the banks and financial institutions). However, it was also recognised that, compared with the social partners, the other stakeholders do not occupy such a central position in relation to leverage with respect to certain debates eg. production methods, duration of working time, pay, other conditions.

II.II ECONOMIC CONTEXT

The Round Table recognised that real economic costs result from the change in production and the re-direction of development necessary to secure sustainable development. The trade-offs between the extent of these costs and the achievement of more environmentally sustainable development represent hard choices which will be evaluated differently by different stakeholders in different situations. Furthermore, the point in the economic cycle at which the evaluations are made will be significant in determining the judgements made. At a time of economic recession the significance attached to the economic costs increases because of a concern with the short-term economic impacts and a reduction in confidence that subsequent economic growth will be large enough to finance the process of change.

The importance of this recognition at the Round Table lies in a more realistic assessment of the possibilities and the timing of achieving sustainable development. Too often, glib and superficial pronouncements on the objectives of sustainability disguise the difficulties, and imply a smooth, swift, cost-free adjustment. As a consequence the expression of the real character of the required change, and hence the necessary development of solutions, has been frustrated. The Round Table, by clarifying the difficulties, has to some extent reduced frustration, evident in some quarters, at a perceived lack of progress towards sustainable development.

The identification of potential economic costs as a major inhibitor to change is also vital to determining the significance of measures and approaches which address directly questions of financial support to industry to manage the process of change and to communities to compensate for the loss of economic activity and employment opportunities. The Round Table has raised serious questions over the efficiency and practicality of the "polluter pays principle", as currently defined and interpreted. For example, the principle as interpreted does not extend to include the social costs associated with adjustments resulting from the internalisation of environmental costs in the costs of production.

II.III OPERATIONALISING SUSTAINABILITY

The Round Table was asked to focus more on the processes and models of change, based upon a broad continuum: from defining issues of sustainability, through a process of defining the interests of stakeholders and the scope for conflict and consensus, to establishing ways of agreeing on co-operation (models of co-operation) between stakeholders in pursuit of sustainability. The early stages in the process of policy development inevitably focus on defining the issues. However, it must be recognised that only through a collective acceptance of responsibility and willingness to participate in the process of change will the process be managed and change implemented.

The continuum gives equal weight to the need to define the central issues **and** the necessary models of co-operation around which the pursuit of sustainability can be developed. The Round Table identified these central issues as including:

- the importance of decentralising actions and the process of change to the local level;

- the importance of industrial relations to the definition of company/sector problems and solutions;

- the importance of engaging SMEs and regionalised industry in the process of change;

- the importance of securing change across the Community, Eastern Europe and Developing Countries.

These issues give rise to separate but overlapping constellations of interests, which can be defined as different starting points, in the development of models. The success of the Round Table lies in the fresh realisation that not only are a range of models required but that some shape to the possible range of approaches can already be defined. The shape derives both from the clarification of the multi-dimensional and economic contexts.

Multi-Dimensional Issues

The various dimensions which need to be recognised in new approaches include:

- the requirement for consistent and coherent supra-national, national and local frameworks embracing different concepts of industry, sustainability and social dialogue; and

- the varying range of stakeholder interests, including those of the social partners, which are likely to differ according to the nature of particular situations.

Round Table on Industry, Social Dialogue and Sustainability

Economic Issues

The broader economic issues which need to be recognised include:

- the requirement for local (especially financial) support for industrial change organised within a supra-national framework; and

- the importance of environmental education and training for all stakeholders.

II.IV NEW MODELS OF CO-OPERATION

Operationalising sustainability requires new models of co-operation. The broad shape of the conceptual framework has begun to emerge, based upon the four dimensions above. The Round Table was presented with two new models of co-operation: one which addressed the issues of support to SMEs to secure improvements in environmental performance, and one which addressed the development of industrial relations, also as an aid to improving the environmental performance of industry.

These models, which are described in more detail in the report, provide a basis from which to develop new ways of recognising and responding to the main dimensions. The Round Table provided clear advice as to the issues which the models need to address and suggestions for their development. This discussion has enabled some appreciation of the necessary developments which are required if the models are to help in operationalising sustainability, by recognising and responding to the main dimensions. For example, the role of education and training was recognised as being vital not only to the attainment of the necessary skills of the social partners to manage change, but also as a fundamental mechanism for developing and defining the interests of all stakeholders and hence to ensuring their necessary involvement in the process of change.

The discussion of these models also highlighted the importance of recognising the inter-relationship between social and productive relations, issues of health and safety and environmental quality. The importance of the relationship was felt to be the different perceptions which emerge from the overlapping interests of different stakeholders as to environmental objectives and the consequent specification of measures for securing objectives. For example, issues of occupational health and safety relate to one set of interests of a particular group of stakeholders and a particular set of environmental objectives, whilst issues of, say, global warming relate to a different set of stakeholders, with different environmental objectives.

III. ROLE OF THE EUROPEAN FOUNDATION

The Round Table was designed, in part, to provide guidance to the European Foundation in the development of its research programme and policy framework. A number of conclusions and suggestions arose from discussions.

- The establishment of a harmonised Community framework within which to develop local and regional initiatives was accepted as an obvious starting point. The recognition of the importance of the wider context to improving living and working conditions at the level of the enterprise should be reflected in attempts to define this wider context (eg. the inter-relationship between supra-natural frameworks such as eco-auditing procedures and enterprise activity). Thus the Commission and Member States are required to ensure that harmonisation proposals reflect and respond to the needs and aspirations of stakeholders at enterprise and community level. At the same time stakeholders at a local level need to be confident that there is a "level playing field", ie. that common objectives of change are shared and owned by the stakeholders (common goals).

- The realisation of the breadth of interest of stakeholders, in addition to the social partners, in securing sustainable development means that the European Foundation has to develop relationships with, for example, representatives of consumer interests. It also has to consider sustainable development not only in relation to the different economic and environmental conditions which prevail within the European Community but also in the context of the problems and opportunities facing Eastern Europe and Developing Countries. The latter countries start from a lower base in terms of the existing structural and institutional framework and hence potentially more radical steps are required to develop models of co-operation. Lessons from the Rio Conference in relation to the issues of global sustainability and development have therefore to be understood within a development of Community frameworks.

- The importance of the development of new models was highlighted. The innovative research of the European Foundation, focusing on industrial relations, the firm in its environment and environmental education and training, represents an important first step in defining these new models. This research should be strengthened and developed. In so doing the interests of different stakeholders needs to be a more explicit consideration.

Round Table on Industry, Social Dialogue and Sustainability

MAIN PROCEEDINGS OF THE ROUND TABLE

The summary of the main issues raised by the Round Table, presented above, draws upon the discussion and debate of the conference. The main proceedings arising from the debate are presented below, using the same structure as the Round Table. The programme of the Round Table and list of participants is presented as Annex A.

There were four themes. The reporting of each theme is structured around the central concepts of industry, social dialogue and sustainability. The themes are:

A. **The Firm in its Local Environment;**

B. **Environmental Education and Training;**

C. **Environment and Industrial Relations; and**

D. **Eco-auditing.**

A final section, **E.**, summarises the main points arising from the concluding panel discussion of the Round Table.

A number of the issues, reported in this document, are the subject of a more comprehensive and detailed treatment in published material, available from the European Foundation. Key publications are:

- The Firm in it's Local Environment : A Consolidated Report.

- The Education and Training of Personnel Concerned with Environmental Issues Relating to Industry.

- Industrial Relations and the Environment.

In addition, a number of international agencies are actively working on the issues raised by the Round Table. These include:

- The Organisation for Economic Co-operation and Development (OECD).

- The International Labour Organisation (ILO).

Round Table on Industry, Social Dialogue and Sustainability

- The United Nations Environmental Programme (UNEP).

- The International Chamber of Commerce (ICC).

These agencies have all produced relevant publications. Interested readers are encouraged to refer to these agencies and the European Foundation for publications.

Round Table on Industry, Social Dialogue and Sustainability

THEME A. THE FIRM IN ITS LOCAL ENVIRONMENT

THE RESEARCH PROGRAMME

The growing demand for improvements in environmental quality and the removal of environmental degradation raises the practical question of how such change is to be introduced and brought about. In terms of environmental pollution caused by industry the central issue is how the environmental performance of small and medium sized enterprises (SMEs) is to be improved, given that large firms have both the awareness and resources to instigate programmes designed to improve their environmental performance.

The obstacles to achieving changes in SMEs are widely known relating in broad terms to infrastructure failure and resource availability. One way to engage SMEs is to involve them in local networks of similar companies facing the same problems and by using these networks to provide information, advice, specialist expertise and financial resources as the basis for co-operation between firms, regulators, and environmental experts.

The research programme sought to identify examples of these networks of co-operation and to review their main characteristics. Some twenty examples were identified and investigated throughout Europe. Summary details are given in Table 1. The consolidated report on the research is now available.

MAIN FINDINGS

The findings of the research encourage the view that local initiatives, which develop co-operation between firms, public authorities and environmental expertise are capable of bringing about positive changes in the environmental performance of SMEs. The features which are important in developing these initiatives include:

- clear objectives;
- strong direction/management/ownership of the initiative;
- social dialogue and participation of existing networks;
- willingness to invite, and follow, specialist advice and expertise.

These initiatives, whilst developing some common features, derive from many different circumstances and are strongly influenced by national characteristics, such as the relative strength and activity of local Chambers of Commerce, or the existence of regional as well as local authorities. However, despite widely differing national characterisations, giving rise to different models, local initiatives of co-operation and social dialogue have developed with the same broad environmental objectives.

TABLE 1 : A SUMMARY OF THE CO-OPERATION INITIATIVES INCLUDED IN THE STUDY

CO-OPERATION DETAILS

Ref No	Name	Region	Brief Description	Scale (No of firms involved
G1	Abfallberatungsagentur (ABAG)	Badden-Wuttenburg, G	• Waste management agency led, with technical experts. • Focus on waste minimisation. • Small and medium sized firms.	30-35
G2	Förderkreis Umweltschutz Unterfranken (FUU)	Unterfranken, G	• Club of members. • Focus on general information exchange. • All firm types but mainly SMEs.	c50
G3	Innovations und Koordinersungsstelle fur die Metallindustrie (IKS)	Ruhr Valley, G	• Innovation and co-ordination agency with strong trade union and worker involvement. • Focus on production and product innovation. • Environmental improvement dimension. • Firms in the metal and electronic sectors, mainly SMEs.	c60
G4	Landesgewerbeanstalt Bayern (LGA)	Bavaria, G	• Technical Institute led. • Focus on technical advice. • All industrial sectors, only SMEs.	c500
UK1	Coventry Pollution Prevention Panel	Coventry, West Midlands, UK	• Long established, multi-party co-operation. • Focus on industrial pollution control. • All firm types included.	c120

Round Table on Industry, Social Dialogue and Sustainability

Ref No	Name	Region	Brief Description	Scale (No of firms involved
UK2	Solent Industrial Environment Association	Southampton, South-East, UK	• Newly established multi-party co-operation. • Focus on awareness raising. • All firm types, but mainly larger firms.	c20
UK3	IT CERES (Southampton University)	Southampton, South-East, UK	• Commett programme funded network. • Focus on environmental training. • Large company orientation.	c12
UK4	The Sheffield Green Business Club	Sheffield, Yorkshire and Humberside, UK	• Planned multi-party co-operation. • Focus on awareness raising. • All firm types included.	not yet known (c50-100)
NL1	Stimulering Afval-En Emissiepreventje Rotteram en Omstreken (Stimular)	Rotterdam, NL	• Innovation department led (regional govt). • Focus on waste reduction. • Aims to bridge the gap between R+D and industry. • Sectoral focus. Only SMEs.	6 pilot projects will be greatly increased
NL2	Preparation of Soil Inventorisation for Industrial Sites, South Holland (BSB)	South Holland, NL	• Multi-party co-operation led by the Chamber of Commerce. • Focus on soil contamination of industrial sites. • All sectors, but only SMEs.	9 pilot projects, will be greatly increased
NL3	Environmental Care in Industry	Gelderland, NL	• Pilot multi-party co-operation led by the Chamber of Commerce. • Focus on environmental management and training. • All sectors but mainly SMEs.	20 firms

Round Table on Industry, Social Dialogue and Sustainability

Ref No	Name	Region	Brief Description	Scale (No of firms involved
F1	Energy Enhancement Unit, Mazamet	Tarn Midi Pyrenees, F	• Chamber of Commerce led industry initiative. • Provides a special facility for the treatment of sludges, to be used collectively. • For SMEs in the wool and leather industry.	60
F2	Organic Waste Treatment Centre, Chateaurand	Bouches du Rhone, Provence-Alpes-Cote D'azur, F.	• Joint public/private venture. • Provides a special facility for the treatment of organic waste. • Service to SMEs, particularly agricultural.	3 directly in the venture, c200 users.
F3	CYPRES	Bouche du Rhone, Provence-Alpe-Cote D'azur, F.	• Joint public/private initiative. • Focus on information exchange. • All industrial sectors and firms, but emphasis on larger firms subject to Seveso regulations.	35 subject to Seveso regulations.
F4	APORA (Employers Anti-Pollution Association)	Rhone-Alpes, F.	• Long established, multi-party co-operation. • Focus on information/advisory services to members. • All industries and firms included.	n/a
F5	CODLOR	Lorraine, F	• Chamber of Commerce led initiative • Focus on information/advisory service to clients (mainly free). • All industries, mainly SMEs.	n/a
S1	AEMA (Spanish Environment Association)	Barcelona, Catalonia, S (and elsewhere)	• Public sector initiative. • Focus on pollution inspection and technical advice. • Focus on companies who have undertaken state aided renovation/extension of plant.	n/a

Ref No	Name	Region	Brief Description	Scale (No of firms involved
S2	VAPSA	Valencia	• Joint public/private venture. • Provides a specialist facility for the treatment of toxic wastes for surface treatment processes. • Surface-treatment industry.	18
S3	BRISA	Madrid, S	• Co-operation of Ministry of Industry and the Association of Manufacturers of Capital Goods. • Focus on information services to members (at a price). • Designed mainly for SMEs.	n/a
S4	METROPOLIS	Bilbao.	• Multi-party co-operation. • Strategic plan for revitalisation. • Range of environmental initiatives for all sectors and firms.	n/a
I1	Industrial Waste Exchange	Turin, Piedmont, I	• Public sector initiative. • Focus on information exchange of wastes. • All firms.	n/a
I2	Regional Society for the Environment	Liguria, I	• Public sector initiatives. • Focus on technical advice. • All industries and firms.	n/a

ROUND TABLE DISCUSSION

The general principles of social dialogue, local and regional development and involvement of local networks were judged by participants to be intuitively a prime mechanism for improving environmental performance in industry. The Round Table emphasised the implications of this approach.

Industry

The approach is relevant throughout the Community but is likely to have most appeal in Member States which have a decentralised pattern of government and where regulatory authorities are able to become involved without compromising their statutory responsibilities. However, the requirement, in many initiatives, for technical advice and technological awareness places a constraint on the approach in regions which lack access to technological resources.

The approach has most appeal where industrial sectors and firms have experience of, and a willingness to, enter into collaborative activity with other firms. Where competitive rivalry inhibits this process then the potential of the initiatives may be constrained. This issue is particularly important for initiatives which adopt a very specialised and sector specific character.

The participation of industry in these initiatives will only occur if firms perceive long-term competitive advantage from doing so. Thus the commercial as well as environmental advantage of participation needs to be highlighted. The consumer, reflecting values of society, demanding better environmental performance represents a fundamental influence in shaping the direct relationship between commercial and environmental performance.

Social Dialogue

The approach explicitly requires, and builds on, social dialogue. However, conventionally defined dialogue, between employers and unions, does not currently represent the central channel of communication. The different models of co-operation exemplify different dialogues developing between a wide range of stakeholders, not just the social partners of employers and workers.

To the extent that a wide range of stakeholders could potentially be involved, the various roles played by stakeholders needs to be considered to prompt debate. For example, the role of banks and financial interests which have hitherto not received much attention may, in future, have to define new arrangements and funding criteria for firms involved in co-operation initiatives and thus exert a significant influence on the character of social dialogue.

Sustainability

The requirement of future environmental policies to rely upon voluntary actions and self-regulation, puts an emphasis on this type of approach. However, the issue of funding actions by firms, who are otherwise unable to finance the necessary time and investment, remains central to a more widespread evaluation of initiatives and extensive participation in such initiatives by SMEs.

To the extent that resource constraints exist but support for the approach builds up, longer-term strategic policy frameworks for regions and local environments are required to ensure that local initiatives focus on issues which have either the greatest direct impact on the environment or have the greatest influence on local firms. Such frameworks are also essential to establishing the inter-relationship between regional environmental and economic development objectives.

As part of this wider framework the institutional roles, particularly of the regulatory agencies, need to be examined to ensure that channels of information and advice are not obstructed by the need to enforce environmental regulations. Social dialogue between institutions and other stakeholders are central to the definition and re-definition of these roles and to define the appropriate political strategies as part of the wider framework.

THEME B. EDUCATION AND TRAINING

THE RESEARCH PROGRAMME

The European Foundation, recognising at an early stage the inter-relationship between health and safety issues and environmental pollution, has undertaken a number of studies of the requirement for environmental training which addresses both sets of issues. For example, training for hazardous waste management responds to both issues and has been the subject of research for several years.

The requirement for training for personnel in industry with responsibility for environmental protection, many of whom also have responsibility for health and safety at plant level, was clear from early research. Less evident was the character of training required and how increased levels of training were to be encouraged.

The requirement to define training needs raises important questions about the necessary skills and competences which are necessary for the better management of the environmental performance of industry. Thus the European Foundation has been actively involved with the European Centre for the Development of Vocational Training (CEDEFOP) to examine more closely the skill requirements of personnel as a contribution to the definition of training requirements.

The research programme has been closely linked to the Firm in its Local Environment and the research into Industrial Relations since issues of education and training, and in particular, the take-up of training, are germane to both areas of research.

MAIN FINDINGS

There are many rationales for environmental education and training. Education and training seeks to:

- reduce the shortage of trained personnel;
- improve the level of skills and competences;
- support the effective implementation of environmental policy;
- respond to changing environmental problems and priorities;
- respond to the need to develop environmental legislation and policies;
- encourage environmentally sensitive management of natural resources;
- respond to the greater integration and multi-disciplinary approaches to environmental management;
- promote the diffusion of environmental knowledge, technology and management throughout the Community;

- encourage greater environmental awareness among the Communities citizens, industrialists and policy makers;
- help to improve the effectiveness of regional development instruments;
- help to realise employment potential in the environmental field.

Thus education and training is needed not only to ensure better environmental performance in industry and hence improve the implementation of environmental policies, but also to facilitate sustainable development. For example, by ensuring that economic development advantages are maximised, aided in part by the impetus education and training provides to technical innovations and technological development, and by ensuring cultural change supportive of greater environmental responsibilities. Hence education and training has a fundamental role to play.

However, the research identified a lack of demand for education and training by industry. The need for training, as perceived by policy analysts was not reflected in the expressed demand for training by industry. The reasons for this relate in large part to a lack of awareness on the part of industry of the importance of developing environmental management competencies, combined with an unwillingness to commit resources.

The result of the failure to demand training has been an inadequate development of suitable training resources. Whilst recent research evidence suggests that the situation is improving with a significant expansion in both training and advisory services, the fundamental importance of education and training means that any remaining inadequacy must be quickly dealt with. Consequently, issues of transferring "good practice" between firms, industries and countries, represents part of the solution and have emerged as central issues for future research.

The research findings of the CEDEFOP studies relate to the process of defining the specification requirements for skills and competences, at all levels of the workforce, to manage and promote the environmental performance of the firms. The research has defined some 78 professional profiles in three different countries (Italy, Germany, UK). This analysis revealed marked differences in the formal and informal definition of environmental responsibilities within the workforce and subsequent training requirements. Nevertheless, the research identified not dissimilar progress in the process of defining professional profiles. Moreover, the common issues raised in each of the Member States as a consequence of developing profiles, provide additional weight to the importance of a greater international exchange of information and experience.

ROUND TABLE DISCUSSION

The pivotal role of education and training was widely discussed and understood by participants and a number of issues arose from debate which elaborate a number of the main findings and point to key issues in the future development of environmental education and training.

Industry

The immediate problem of a lack of awareness and motivation was recognised. The perceived costs of developing environmental management responses were argued to be a major constraint. One issue raised was the need to highlight the resource savings from improved management which provides a contribution to training and consultancy costs, as an important means in raising the motivation of industry to respond to defined training needs.

The "chicken and egg" situation was highlighted where lack of definition of environmental management skills reflected a limited appreciation of necessary management responses, itself reflecting a lack of awareness of the need for change. In this context a number of participants raised the basic role of education in society as a vehicle for raising and maintaining environmental values and responsibilities. Without the appropriate cultural conditions changes in the allocation of resources by firms to improve environmental performance were likely to be limited.

As well as cultural change, resulting in part from education, the redefinition of appropriate industry skills can benefit from the existing process for the definition of skills to manage health and safety issues. The strong overlap in many firms between health and safety and environmental performance provides opportunities for the definition of new skills. Given that this overlap provides for communication between employers and workers, and between firms and public authorities, particular benefits can be expected if the opportunity is taken to expand the terms of reference of health and safety committees, etc to consider environmental management issues.

Social Dialogue

The importance of social dialogue in relation to the development of education and training is crucial. Firstly, dialogue in firms between employers and workers is vital to the redefinition of skill requirements and ultimately changes in personnel profiles. Secondly, dialogue between firms and regulatory/public authorities has a fundamental role in developing a clear idea of environmental responsibilities and providing direct channels of communication of information and advice. Furthermore, dialogue is important in defining public programmes of vocational education and training targeted at improving environmental management skills. Thirdly, dialogue between firms within industrial sectors also represents a means of defining new skill requirements as does dialogue between firms and suppliers of environmental protection goods and services.

However, whilst dialogue in various forms can increase the level of environmental education and training, the most important relationship especially in the long-term, is the contribution of education and training to the quality and volume of social dialogue. The most fundamental role is the development of long-term cultural norms which value environmental quality and better environmental performance.

Round Table on Industry, Social Dialogue and Sustainability

Sustainability

The central message, in this context, from the Round Table was the underpinning which education provides to the cultural values of society in favour of environmental responsibility. In the long-term, without this cultural value system there is no understanding of the need for environmental management responses and no willingness to invest in the necessary changes.

The long-term requirement of sustainability of gradually harmonising the environmental performance between industries and countries emphasised a feature of potential future education and training; namely the common definition of skills and competences which ensure a comparable level of environmental performance.

THEME C. INDUSTRIAL RELATIONS AND THE ENVIRONMENT

THE RESEARCH PROGRAMME

The Background Paper identified a distinction between the issues relating to the external relationships which firms have and which impact on the corporate responses to environmental pressures and internal relationships. The contribution of internal relationships to the development of environmental performance includes contributions to internal management and accounting systems and process/product development. A central element of any internal relationship is the dialogue between employers and workers as part of the development of corporate industrial relations.

The potential importance of industrial relations to the shaping of environmental management responses has hitherto been largely neglected in studies of environmental performance. However, in principle industrial relations are a potentially rich source of ideas for better environmental performance (including the specification of skills requirements) as well as providing tried and tested channels of communication to facilitate corporate and sectoral change. Commensurate with the European Foundation's interest in the shaping of living and working conditions, a research programme to investigate the role of industrial relations in developing company responses has been introduced.

The research programme was set up in 1990 and has investigated the development of industrial relations vis-a-vis environmental performance in each of the EC Member States identifying the roles played by the Social Partners and the contribution which industrial relations can make to the implementation of environmental policy. The research is not yet complete but interim policy findings are available.

THE MAIN FINDINGS

The research has identified several distinct phases of development in the manner in which firms have responded to environmental pressures:

- Stage 1 : Employers and Unions unite in resisting environmental pressures;

- Stage 2 : Limited acceptance by employers of pressures reflected in real and presentational differences in corporate performance. No, or limited, role for industrial relations;

- Stage 3 : Self-organisation involving the development of environmental strategies and the full mobilisation of corporate resources to improve environmental performance. Increasing role for industrial relations;

- Stage 4 : Evaluation and confirmation of environmental strategies, designed to establish the environmental improvement attributable to the strategies as well as the corporate impact. Explicit utilisation of industrial relations as an element of the evaluation.

The research suggests that most firms are located somewhere between Stages 2 and 3.

To the extent that the legitimacy of industrial relations derives from legislation (for example, in the case of health and safety issues) then the research suggests that environmental legislation has hardly affected industrial relations. However, in so far as environmental legislation provides a basis upon which trade unions can demand participation rights, the situation could change. These participation rights include access to information and training, and participation in consultation exercises. Potentially, they could extend to the acceptance of co-determination in corporate environmental policies. The evolution of participation rights is partly a function of domestic and traditional and sectoral evolution of industrial relations and thus the pace with which firms move to Stage 4 will differ between firms, sectors and countries.

Research has shown that motives for expanding industrial relations to include environmental affairs include the interest of trade unions in terms of health and safety and job preservation, and as citizens affected by environmental problems. Employers perceive advantage from avoiding conflict and from securing the active involvement in, and commitment to, corporate policy.

Investigation of existing, formal agreements between employers and trade-unions which guarantee participation in corporate environmental affairs indicates the extreme rarity of such agreements at anything other than at plant levels. A few sectoral (eg chemicals) and regional (eg in Italy) agreements have been reached. Plant level agreements which exist, are mainly based upon the acceptance of extending dialogue on internal environmental interests, in health and safety, to issues relating to the local working environment adjacent to the plant.

Despite these agreements the research indicates that the vast majority of employers are of the view that environmental management is not an issue for industrial relations.

ROUND TABLE DISCUSSION

The introduction of industrial relations as a potent force for the development of better environmental performance was well received by participants. This acceptance was based in large part upon the view that legitimate concerns with health and safety encompassed, in principle, environmental issues, at least at local level. However, it was not unexpected that very few examples of co-operation between employers and trade unions extended in this area.

Industry

The discussion raised the issue of distinguishing between formal agreements in the context of established industrial relations mechanisms and voluntary co-operation between employers and workers. The absence of formal agreements does not necessarily mean that concrete proposals and projects for improving environmental performance can not be shaped by social dialogue. However, it is conceivable that the full participation by workers is unlikely without a formal framework, particularly in larger firms. The absence of agreements may mean that co-operation does not become generally accepted and binding.

Moreover, in the context of informal industry agreements, the absence of guidelines may mean that co-operation is less productive. This is a particular issue for SMEs when formalised industrial relations are less in evidence and where voluntary co-operation is more likely.

Participants also made the point that, in so far as formal relationships may be used to ensure job preservation above environmental interests, that agreement may not necessarily be productive in developing corporate responses. Where employment protection represents a major issue then the internal social dialogue needs to expand to include external dialogue which encompasses national and local authorities, trade associations, etc which may be able to assist in the adjustment process by safeguarding employment. The importance of a social "safety-net" or "parachute" was raised by a number of participants, in both this and other contexts.

Social Dialogue

The discussion responded to the importance of defining the appropriate interest of workers, in environmental objectives, be it localised or global. The research has recognised that environmental objectives exist in multiple forms. Figure 1, presented at the Round Table, distinguishes five, overlapping, areas of environmental interest.

Health and safety, as noted above, is conventionally considered as part of industrial relations. This interest could be expanded to include an interest in the quality of the environment inside the plant where it does not represent a direct health or safety issue, and extended still further to an interest in environmental protection due to the risks associated with the production processes and/or products of the plant. The interest of workers in the quality of the local environment, perhaps attributable to their plants, but not necessarily, represents an important community as well as labour interest. Interests in global environmental protection reflecting labour interests as concerned citizens of the world, represents the final area of concern.

These overlapping interests serve to highlight the potential for different perceptions and interpretations of what is meant by environmental performance and what constitutes

FIGURE 1 : LINKS BETWEEN LABOUR INTERESTS AND ENVIRONMENTAL INTERESTS

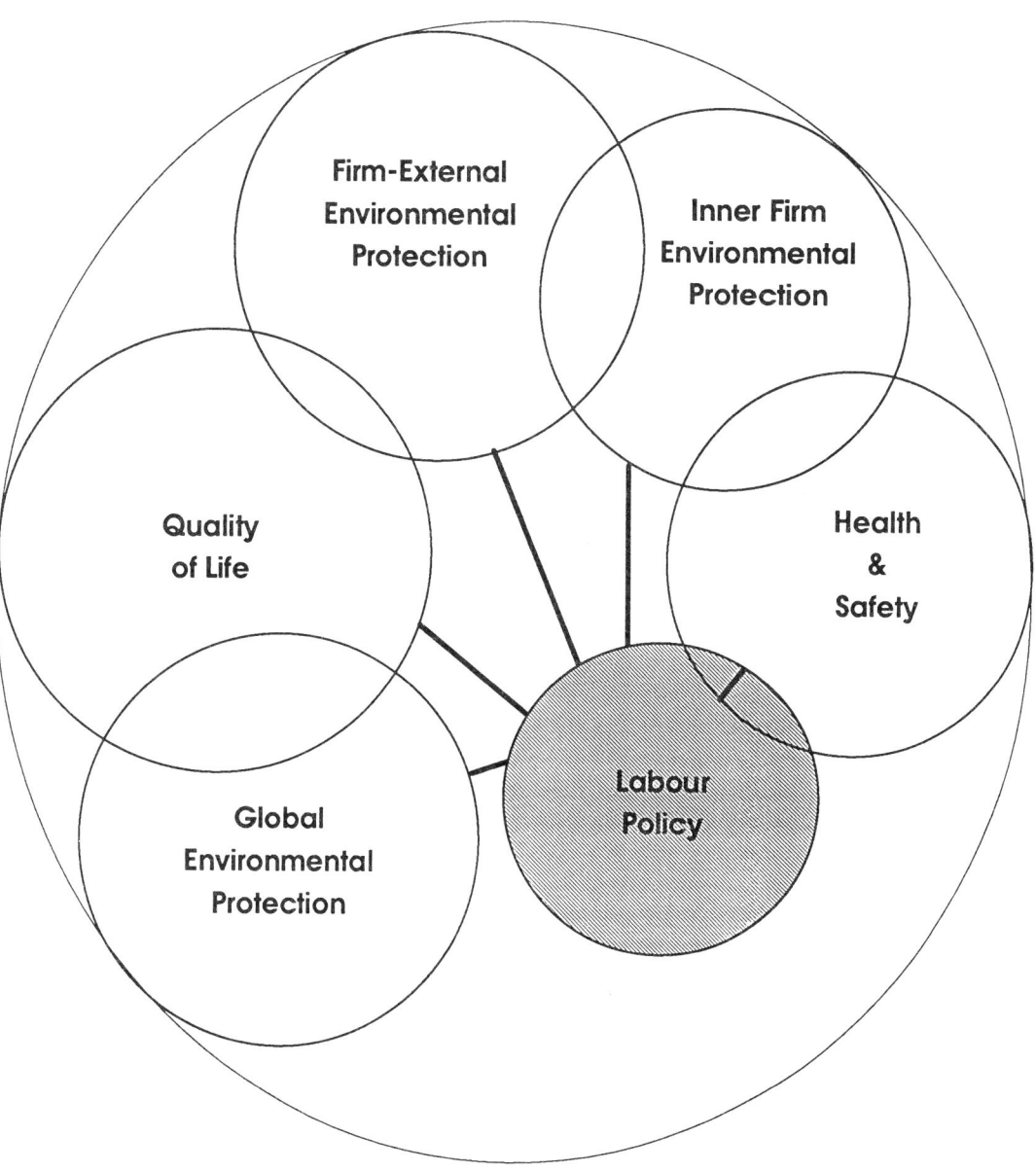

Source : Hildebrandt & Schmidt, Industrial Relations and the Environment Network Europe (IRENE). Research Findings for the European Foundation for the Improvement of Living and Working Conditions, 1992

improvements in corporate behaviour. It is not yet clear which of these particular interests are likely to predominate in subsequent agreements.

The definition of environmental objectives also raised the issue of whether conventionally defined social dialogue and the role of the social partners would change to reflect the accepted collective view of appropriate environmental interests. For example, expansion of industrial relations to include quality of life concerns may mean that representation from the local community, who are not workers at the plant, may become involved in the dialogue. The range of stakeholders could expand beyond simply employers and workers.

Sustainability

The potential long-term importance of revised industrial relations, allowing a more constructive and broader approach to environmental performance were recognised. A key question is what circumstances might be likely to influence the required expansion of industrial relations. The Round Table considered a number of features including:

- the extent of accepted environmental objectives, based upon agreed definitions of environmental interests, and hence the pressure on actors to secure change;

- the extent of wider environmental pressures and values supporting change;

- the capacity of the original corporate framework to adjust, based upon the perceived economic potential and capacity for innovation and planning resulting from developing industrial relations. To the extent that this capacity is a function of the experience of social dialogue then certain sectors (eg. chemicals) may already demonstrate a stronger potential to adjust.

Finally, it was noted that some parallels were emerging from discussion between the firm in its environment and the environment and industrial relations. Three commonalities were identified. Firstly, the importance of defining and accepting common environmental objectives. Secondly, the importance of recognising the breadth of social dialogue and the range of potential stakeholders. Thirdly, the underlying importance of cultural values to sustain the process of change.

Round Table on Industry, Social Dialogue and Sustainability

THEME D. ECO-AUDITING

THE PROPOSAL

The European Commission has proposed a Council Regulation establishing a Community scheme for the evaluation and improvement of environmental performance and the provision of relevant information to the public. This proposal, commonly termed the Eco-audit regulation will, if enacted, provide a formal framework within which firms can undertake Eco-auditing activities throughout the European Community. In so far as it complements existing international (eg. ISO 9000) standards the proposal contributes to a global framework for establishing the quality of corporate activity in implementing environmental policies.

Details of the proposal have been published (see the Official Journal, December 1991). The main characteristics, as introduced to the Round Table, are summarised below. The intention is to experiment with the proposal and establish it's strengths and weaknesses in the context, primarily, of manufacturing industry in the next four years, before seeking a more widespread application.

Objectives of the Proposal

The Regulation seeks to promote a better environmental performance in the framework of industrial activities by:

- encouraging industries to establish and apply environmental management policies, programmes and systems for their sites;

- by allowing auditors and the like to conduct routine, objective and periodic assessments of the effectiveness of these factors;

- publishing each year "environmental declarations" on the level of environmental performance achieved by the firms and its conduct towards the environment.

The formulation of the proposal as a Regulation is important because it means that, if enacted, Member States are required to set up the necessary accreditation system. These systems are charged with setting and implementing appropriate conditions and procedures for the accreditation of environmental verifiers and of supervising their activities. These verifiers are responsible for ensuring that eco-audits are conducted in compliance with the conditions laid down in the Regulations and professional codes of practice to be established by the Commission. The Member State is responsible for establishing and updating a list of accredited environmental verifiers.

Whilst the Regulation determines that an accreditation system is set-up and lays down the criteria both in relation to the conduct of an eco-audit and its verification the participation of companies is voluntary. Firms which volunteer to take part are required to:

- carry out an on-site environmental analysis and use the findings to establish an environment programme for the site, as well as an environmental management system applicable to all the company activities;

- carry out or commission someone to carry out eco-audits on the site n question, in accordance with certain criteria (see Table 2) and in keeping with certain requirements concerning the approval of the environmental auditors;

- set out objectives at the highest managerial level so as to ensure a constant improvement in a firms environmental performance;

- draw up a specific environmental declaration, for public information purposes, on each site covered by an audit;

- have the policy, programme, management system, the audit procedure and the environmental declarations examined;

- have the environmental declarations validated by the competent authorities;

- forward the validated environmental declaration to the national body competent for the site and distribute it, if need be, for inspection by the general public in the country concerned, after the site in question has been registered.

Firms choosing to participate in the scheme will undertake to establish "environmental protection systems" within which the environmental management of the firms can be implemented and audited.

Such systems shall include the following elements, formally stated in writing:

- an environmental policy;
- environmental objectives and targets;
- an environmental programme;
- an environmental management system, including an audit programme.

The documents shall explain how the environmental policy and the environmental management system for the site relate to the policy and systems of the company as a whole, and shall include a statement of the environmental policy for the company.

Within the framework of the environmental protection system shall be set, in particular, technical and organisation measures and procedures aiming at generating the information and data necessary to evaluate, with reference to company's environmental policies, objectives and programmes, the environmental performance of activities in the site considered.

THE IMPLICATIONS OF THE PROPOSAL

Industry

Membership of the eco-audit system will be beneficial for a company's image. First of all, the list of a sites recorded in the Community would be published each year by the Commission in the EC Official Journal. Second, companies would be entitled to use, for their registered sites, "membership declarations" possibly in the form of a logo, whose aim would be to set out clearly the nature of the system. These declarations could not be used to publicise a company's products or be affixed to the products themselves or the packaging.

The benefits of eco-auditing extend well beyond those associated with the publicity attached to public declaration. Environmental performance improvements are often consequent upon changes in corporate activity which is in itself commercially attractive. Paybacks on environmental investment can be very short.

However, it is clear that the proposal requires investment in developing the environmental management system and the auditing process. This has implications for SMEs who are less likely to have already established relevant systems and who could benefit substantially from the system. The encouragement of SMEs to participate has received particular attention. One suggestion to reduce the burden on SMEs is to combine the auditing with the audit-verification process.

The eco-auditing process can be undertaken by the firm or it can be undertaken by consultants. However, in either case the auditors need to demonstrate that their competences and the information system are adequate. One of the criterion in the verification process is the reliability of the information and data used. The main issues to be included in the environmental protection system and hence in the eco-audit are summarised in Table 2.

The eco-audit requirements have clear implications for the management and technical skills required if firms are to contribute to, and learn from, the eco-audit process. The development of suitable skills extends however, beyond simple familiarity with auditing procedures but to a more fundamental appreciation of the interaction between corporate activity and the environment and between corporate strategy and detailed policies and actions.

TABLE 2 : ISSUES TO BE CONSIDERED BY THE ECO-AUDIT

- assessment, control and prevention of interactions of corporate activity with the various media of the environment;
- energy management, saving, choice;
- raw materials, management, savings, choice and transportation; water management and savings;
- waste reduction, recycling and reuse, transportation, elimination;
- selection of production processes;
- products planning (design, packaging, transportation, use and elimination);
- prevention and mitigation of accidents;
- staff information, training, participation, on environmental issues;
- external information and participation of the public, including dealing with public complaints.

Social Dialogue

The Round Table discussed in some length the provisions for, and guidance on, the participation of workers and trade-unions in the auditing process. The Regulation leaves the process of corporate response to eco-auditing requirements, within the criteria laid down, to the individual firm. The issue then becomes one of considering how the development and application of industrial relations can improve the process of environmental management and hence environmental performance, reflected in the results of the eco-audit.

The eco-audit has particular implications for other stakeholders. For example, the publication of the eco-audit will allow representatives of communities affected by firms to understand and participate in a dialogue with firms designed to assist environmental performance. To the extent that the eco-audit indicates the presence of good environmental management activities and/or the absence of liabilities associated with environmental pressures, the eco-audit can convey a strong message to stakeholders with a commercial interest in the company.

Sustainability

The eco-audit proposal represents an important stage in establishing a common framework within which to improve corporate environmental performance. By laying down common criteria and enforcing verification systems, the proposal provides both a basis for assessing performance and a confidence to interested parties that the assessment is fair and reasonable. Clearly, it will take time to develop comparable national interpretations of what a fair and reasonable eco-audit should attempt to analyse and to communicate. However, in principle at least, the proposal provides an important tool in the pursuit of better environmental performance by industry, by providing firms with an independent benchmark of performance and stakeholders with information to assess firms performance in relation to their own values.

THEME E. PANEL DISCUSSION

INTRODUCTION

The Round Table provided the opportunity for two panel discussions. The first panel enabled leading international organisations to summarise their overall work in the area of environmental policy and sustainable development. The second panel provided a stronger, European focus on the substantive issues. We synthesise below the main observations resulting from the two panels.

KEY POLICY DIRECTIONS OF THE INTERNATIONAL ORGANISATIONS

The first Panel comprised presentations by representatives of:

- United Nations Environmental Programme (UNEP)
- Organisation for Economic Co-operation and Development (OECD)
- International Chambers of Commerce (ICC)
- International Labour Organisation (ILO)

Figure 2 briefly summarises the main points arising from these presentations. For a fuller understanding of the work of the organisations readers are recommended to make direct contact.

FIGURE 2 : KEY POLICY DIRECTIONS OF THE INTERNATIONAL ORGANISATIONS	
☐ UNEP	• Strategic Policies to Promote Sustainable Development • International Co-operation and Partnership Development • Local Initiatives (technology transfer, education & training, etc)
☐ OECD	• Integrated Development of Economic and Environmental Policies • Developing Dialogue with Industry • Role for Market Mechanisms and Voluntary Accord
☐ ICC	• Promotion of Business Charter for Sustainable Development • International Dialogue and Policy Formulation • Social Dialogue Development for Policy Implementation
☐ ILO	• International Framework for Sustainable Development • Integrated Development of Economic and Environmental Policies • Clearer Evaluation of Economic and Environmental Impacts

The presentations revealed a strong commonality of purpose and direction, in accord with the general conclusions of the Round Table. The strength of the conclusions derive in part from this degree of consensus, between the different international organisations. Common strands included:

- the importance of taking a longer-term view in which the overlapping interests of different stakeholders can be defined and integrated;

- the importance of developing both international programmes and policies as a framework for detailed and specific measures at a local level and highlighting the importance of local initiatives by firms, workers and communities to securing social dialogue and sustainable development;

- the definite and concrete acceptance of the underlying importance of social dialogue in reaching agreements on sustainability;

- the realisation that significant adjustment costs are likely and that subsequent issues of equity and fairness need to be addressed;

- existing examples of co-operation, even between the individual organisations represented at the Round Table;

- the central role of education and training in raising awareness, and facilitating participation in environmental improvements exemplified in the programmes of the different organisations.

The views expressed by the International Panel are also reflected in the following synthesis of the debate of the European Panel.

Industry

The environmental impact of industry is wide ranging. Whilst it is reasonable to focus on particular firms and sectors, especially large manufacturing companies, and certain environmental impacts as a practical way forward, it must be remembered that industry includes a wide range of primary and tertiary economic activity, often in smaller firms, generating a diverse set of environmental impacts.

The need for industrial restructuring and change as a process of adaptation to meet environmental goals and values is not, any longer, the central issue. This is, at least for many governments and social partners, now accepted. The central issue relates to how quick the adaptation process should be and the need to recognise and overcome obstacles to the process of change.

The focus on means, rather than ends, raises, for example, issues relating to the practical application and interpretation of the polluter pays principle. Recognising the importance of introducing market based instruments to complement environmental regulations and legislation, has a subsequent requirement to quantify and value the appropriate environmental impacts. The assumption that the external costs can be identified and then internalised by sending appropriate price signals demands major advances in the evaluation of environmental impacts and the understanding of firms behaviour in response to changes in market conditions.

The process of adaptation by industry raises two basic requirements for firms. Firstly, firms need to be motivated to, and capable of, responding to environmental pressures. The fundamental role of education and training to both sustain the values which underpin motivational and behavioural changes and to provide the necessary managerial and technical skills can not be over emphasised.

Secondly, in many instances adaptation not only requires new attitudes and competences but also the necessary technical solutions to problems which derive from the production process. The relationship between the process of adaptation to environmental pressures and technological development and change is very strong, reflected in many technology centred activities designed to improve environmental performance. The ready availability of appropriate technology represents a central issue.

The process of change, from the point of view of governments and social partners, needs to be fostered by an appreciation that environmental costs associated with adaptation are commensurate with long-term improvements in competitive advantage, profitability and security of employment. Whilst the relationship may have intuitive appeal a stronger documentation of the impacts associated with the process of change needs to be prepared and diffused.

Social Dialogue

If industry is now in a position where the discussion can focus on means rather than ends then this is the consequence of the success of social dialogue in transforming the environment from being a peripheral and negatively perceived set of issues into the central, long-term challenge and opportunity facing society.

The underpinning process of social dialogue is paramount if environmental objectives are to be continually defined and redefined and if the process of change and response is to be implemented and managed effectively. The Round Table itself is a manifestation of the goodwill and common purpose which lies behind the process of change. However, since the process of change has only just begun (and could conceivably be reversed) the process of social dialogue must continue.

A number of common messages were presented:

- the sensitisation of consumers and the public to environmental issues and the encouragement of environmental values lies at the heart of the process of social dialogue;

- the role of education and training has therefore a twin objective: to facilitate this sensitisation and to provide the necessary motivation and competences to manage the process of change. Social dialogue is imperative if these objectives are to be fulfilled;

- the responses of firms have to be assisted by revisions and adaptations to internal company relationships. The conduct of industrial relations as a central dialogue has the potential to play a fundamental role in both influencing values and attitudes and contributing to practical initiatives for change;

- the obstacles of information failure and finance, in the path of change, are real. Social dialogue is fundamental to identifying and addressing these obstacles;

- the process of social dialogue is fundamental to achieving a global, rather than a piecemeal and uneven, response. The commonality of issues facing the western world, also arises, with greater intensity, in the developing countries and in eastern Europe. The transfer of experience between different social partners and stakeholders is essential to addressing the issues on a world wide scale.

Sustainability

The pursuit of sustainable economic development will be manifested in a wide variety of different actions, initiatives and policies. The underlying requirement is that governmental and corporate policies which are discrete from environmental policy, should become integrated with environmental policy. The process of integration is necessary at all levels of government and corporate activity.

The process of integration should recognise and respond to the real concerns over the costs of adjustment, at both a corporate and societal level. Failure to respond will lead to inertia and, in some cases, the return to short term policies which rely on resource depletion without regard to the longer term effects. Tackling this issue, which has especial relevance in the Third World, requires the common sense of purpose and acceptance of environmental responsibilities which can only emerge from a healthy and vibrant social dialogue, together with concerted actions to soften, where possible, the impact of change on people's welfare.

Striking the correct balance between protecting the welfare of society now and in the long-term represents the basic challenge facing not only industry but political and social institutions and agencies. Devising and implementing policies and actions to achieve the appropriate balance has begun. The ideas expressed at the Round Table will encourage and expand the process of change.

ANNEX A

**European Round Table on
"Industry, Social Dialogue and
Sustainability"**

The post-Rio instruments and perspectives

Rome, 30 November-1 December 1992

Round Table on Industry, Social Dialogue and Sustainability

PROGRAMME

MONDAY 30 NOVEMBER

08.30	*Registration*
09.00	**OPENING SESSION** Chairman : Jørn PEDERSON, European Foundation for the Improvement of Living and Working Conditions **Opening of the Round Table** by Eric VERBORGH, Deputy Director of the European Foundation for the Improvement of Living and Working Conditions
10.00	Introduction to the issues of the Round Table, by James MEDHURST, ECOTEC, Birmingham
10.15	*Coffee/tea break*
10.45	**FIRST SESSION** Chairman: Nicola MESSINA, FEDERCHEMICI and Carlo Fabio CANAPA, UIL **The Firm in its Local Environment**. Presentation of the Foundation's project by Voula MEGA, European Foundation for the Improvement of Living and Working Conditions, and James MEDHURST, ECOTEC, Birmingham
11.15	Discussion
11.45	**Education and Training regarding the Environment**. Presentation of projects, by Jørn PEDERSON, European Foundation for the Improvement of Living and Working Conditions, and Gesa CHOME, CEDEFOP, Berlin.
12.15	Discussion
13.00	Closure of the first session, by Dr Maurizio Polverari, representative of the Minister for Labour. Lunch

Round Table on Industry, Social Dialogue and Sustainability

14.30	**SECOND SESSION**
	Chairman: Corrado CLINI, Ministry of the Environment
	Industrial Relations and the Environment: The Situation in Europe. Presentation of the Foundation's work, by Kees LE BLANSCH, CEM, The Netherlands, and Eckart HILDEBRANDT, Wissenschaftszentrum, Berlin
15.15	Discussion
15.45	*Coffee break*
16.15	**THIRD SESSION**
	Chairwoman: Anna CARLI, CGIL
	The Eco-auditing approach and Instruments of the Commission. Presentation of the Commission's work and its perspectives by **Bernado Delogu**, Directorate-General XI, Commission of the European Communities
17.00	Discussion
18.00	End of third session.

Round Table on Industry, Social Dialogue and Sustainability

TUESDAY 1 DECEMBER

09.15 **FOURTH SESSION**

Chairman: Eric VERBORGH, Deputy Director, European Foundation for the Improvement of Living and Working Conditions

Panel on International Concern for Sustainability

Participants: Clare DELBRIDGE, UNEP, A.S. BHALLA, ILO, Jeremy EPPEL, OECD and Nigel BLACKBURN, International Chamber of Commerce

11.00 *Coffee break*

11.30 **CLOSING SESSION**

European Panel on Industry, Social Dialogue and Sustainable Development

Moderator: Hugh WILLIAMS, ECOTEC, Birmingham

Participants: Bernado DELOGU, CEC, Giuseppe CACOPARDI, Ministry of Labour, Michel MILLER, ETUC, Serge CHIMKOVITCH, UNICE, Raffaele MORESE, CISL, M BRIGNONE, Confindustria

12.45 Discussion

13.30 Closure of Round Table by Eric VERBORGH

ORGANISING COMMITTEE

Management: **Jørn PEDERSON**
Voula MEGA
Administration: **Ann McDONALD**
Catherine CERF

With the kind collaboration of the Italian members of the Administrative Board of the Foundation:

Cecilia BRIGHI, CISL
Giuseppe CACOPARDI, Ministry of Labour
Raffaele DELVECCHIO, Olivetti

General Rapporteur: **James MEDHURST**, ECOTEC

Venue: **CNEL** (Consiglio Nazionale dell'Economica e del Lavoro)
Viale David Lubin, 2
00196 ROMA
Tel : + 39 6 3692273/3610469

Round Table on Industry, Social Dialogue and Sustainability

LIST OF PARTICIPANTS

Luciano BARBATO
Confederazione Italiana Sindacati
Lavoratori
Unione Sindacale Regionale del Lazio
via Carlo Cattaneo 23
00185 Roma
Italy
Tel : 6-444941
Fax : 6-4465642

A S BHALLA
Chief
Employment Strategies Branch
Employment and Development Department
ILO
4 route des Morillons
CH-1211 Geneva 32
Switzerland
Tel : 79 96 415
Fax : 79 88 685

Nigel BLACKBURN
International Chamber of Commerce
Commission of Environment
38 Cours Albert Ier
75008 Paris
FRANCE
Tel : 1-49 53 28 28
Fax : 1-49 53 28 59

Enrico BONO
FEDERTESSILE
Viale Sarca, 223
20126 Milano
Italy
Tel : 2-66103440
Fax : 2-66103455

Luca BORGOMEO
Confederazione Italiana Sindacati
Laboratori (CISL)
Via Po, 21
00198 Roma
Italy
Tel : 6-84 73 377
Fax : 6-84 73 203

Cecilia BRIGHT
Confederazione Italiana Sindacati
Laboratori (CISL)
Via Po, 21
00198 Roma
Italy
Tel : 6-84 73 377
Fax : 6-84 73 203

Salvatore BUCOLO
CISL
Via Cociferi 55
Catania
Italy

Bruno BURRAI
ASAP
via Due Macelli 66
00187 Roma
Italy
Tel : 6-673 4247
Fax : 6-673 4242

Round Table on Industry, Social Dialogue and Sustainability

Giuseppe CACOPARDI
Direzione Generale dei Rapporti
di Lavoro
Ministero del Lavoro e della
Previdenza Sociale
Via Flavia 6
00187 Roma
Italy
Tel : 6-4482 48 16
Fax :6-461 44 96

Carlo Fabio CANAPA
UIL
Via Lucullo, 6
00198 Roma
Italy
Tel : 47 531
Fax : 47 53280

Anna CARLI
Confederazione Generale Italiana
del Lavoro (CGIL)
Corso d'Italia, 25
00198 Roma
Italy
Tel : 6-847 61
Fax :6-88 456 83

Catherine CERF
European Foundation for the Improvement
of Living and Working Conditions
Loughlinstown House
Shankhill Co.Dublin
Ireland
Tel : 1-2826888
Fax : 1-2826456

Serge CHIMKOVITCH
Industrial and International
Relation Manager
SOLVAY
Rue du Prince Albert, 33
1050 Bruxelles
Belgium
Tel : 32-2 509 61 11
Fax : 32-2 509 72 40

Gesa CHOME
CEDEFOP
Bundesallee 22
1000 Berlin 15
Germany
Tel : 004930-88412164
Fax : 004930-88412222

Benito CIUECI
CISL-Internazionale Lazio
via Carlo Cattaneo 23
00185 Roma
Italy
Tel : 6-44494
Fax : 6-4465642

Corrado CLINI
Direttore Generale
Inquinamento Atmosferico
Acustico e Rischio Industriale
Ministero dell'Ambiente
Via Della Ferratella, 33
00184 Roma
Italy
Tel : 06-7027349
Fax : 06-7027195

Round Table on Industry, Social Dialogue and Sustainability

Carlos COELHO
SEIA
Sociedad de Studios de Impact Ambiental
Rua C Edeficio 125
3e étage
Aeroporto de Lisboa
1700 Lisboa
Portugal
Tel : 01-8475036
Fax : 01-801380

Francisco CORREIA
LNEC
DH
Av. Do Brasil, 101
1799 Lisboa Codex
Portugal
Tel : 1-8482139
Fax : 1-8473845

Ada Lucia DE CESARIS
Ist. Per L'Ambiente
Via Emanueli 15
PO Box 10098
20110 Milano
Italy
Tel : 2-661301
Fax : 2-66102201

Marc DE GREEF
Directeur
ANPAT
Rue Gachard 88
1050 Bruxelles
Belgium
Tel : 2-648 0337
Fax : 2-648 6867

F DEFILIPPI
c/o ITALTEL
via Toqueville 13
Roma
Italy
Tel : 6-43885107

DEL ZOPPO
CONFARTIGIANATO
Divisione Politica Economica
Settore Ambiente
Via San Giovanni in Laterano, 152
00184 Roma
Tel : 6-70452188

Bernard DE POTTER
ACV Dienst Orderneming
Wetstraat 121
1040 Brussels
Belgium
Tel : 2-237 36 39
Fax : 2-233 37 00

Clare DELBRIDGE
Industry and Environment
Programme Activity Centre
Tour Mirabeau
39-43 Quai André Citroen
75739 Paris Cedex 15
France
Tel : 33-1 40 58 88 50
Fax : 33-1 40 58 88 74

Bernardo DELOGU
Directorate-General
Environment, Nuclear Safety
and Civil Protection
Commission of the EC
Rue de la Loi 200
B-1049 Brussels
Belgium
Tel : 299-22-63
Fax : 299-08-95

Raffaele DELVECCHIO
Direzione Relazioni Industriali
Olivetti S.p.A
Via G Jervis, 77
10015 Ivrea (To)
Italy
Tel : 125-522593 Ext. 2593
Fax : 125-522150

Jeremy EPPEL
OECD
Counsellor
Environment Directorate
2 rue André-Pascal
75775 Paris Cedex 16
France
Tel : 1-45 24 79 13
Fax : 1-45 24 78 76

Jean-Pierre GIRAUDON
Chambre Régionale de Commerce
et d'Industrie
BP 158
Avignon Cedex
France
Tel : 33-90 82 40 00
Fax : 33-90 85 56 78

Denis GREGORY
Ruskin College
Dunstan Road
Old Headington
Oxford
OX3 9BZ
United Kingdom
Tel : 0865 750029
Fax : 0865 742195

Franco GIUSTI
Confindustria
FIAT S.p.A.
Relazioni Industriali
Corso Marconi, 10
10125 Torino
Italy
Tel : 011-65652250
Fax : 011-6663771

Eckart HILDEBRANDT
Wissenschaftszentrum Berlin
Reichpietschufer 50
1000 Berlin 30
Germany
Tel : 254 91 279
Fax : 254 91 684

Neal HOGGAR
IVECO FIAT SpA
Via Puglia 35
10156 Torino
Italy
Tel : 22-42 029
Fax : 22-44 500

Round Table on Industry, Social Dialogue and Sustainability

Henrik HOUGS
Confederation of Danish Industries
1787 Copenhagen V
Denmark
Tel : 45-33773377
Fax : 45-33773370

Norbert KLUGE
Hans-Böckler-Stiftung
Abt. Forschungsförderung
Bertha-von-Suttner-Platz 3
4000 Düsseldorf
Germany
Tel : 049-211 7778175
Fax : 049-211 7778120

Hubert KRIEGER
European Foundation for the
Improvement of Living and Working
Conditions
Loughlinstown House
Shankhill Co. Dublin
Ireland
Tel : 1-2826888
Fax : 1-2826456

Kees LE BLANSCH
Centrum voor Beleid & Management
Universiteit van Utrecht
Munstraat 2A
3512 EV Utrecht
The Netherlands
Tel : 030-53 81 01
Fax : 030-53 61 56

Bernard LE MARCHAND
UNICE/CLE
76, avenue v Gilsoul
1200 Brussels
Belgium
Tel/Fax : 2-771 58 71

Børge LORENTZEN
Interdisciplinary Centre
Technical University of Denmark
Building 208
DK-2800 Lyngby
Denmark
Tel : 42 88 22 22 ext. 5952
Fax : 42 88 20 14

Franco LOTITO
UIL
Via Lucullo, 6
00198 Roma
Italy
Tel : 47 531
Fax : 47 53280

Settis MASSIMO
Unione di Torino
Via Fanti 17
10128 Torino
Italy
Tel : 011/57181
Fax : 011/544634

Franco MAZZA
CGIL
Dip.to Ambiente
Corso d'Italia 25
00198 Roma
Italy
Tel : 6 84 76 458
Fax : 6 884 56 83

Ann McDONALD
European Foundation for the Improvement
of Living and Working Conditions
Loughlinstown House
Shankhill Co. Dublin
Ireland
Tel : 1-2826888
Fax : 1-2826456

Voula MEGA
European Foundation for the Improvement
of Living and Working Conditions
Loughlinstown House
Shankhill Co. Dublin
Ireland
Tel : 1-2826888
Fax : 1-2826456

James MEDHURST
ECOTEC Research and Consulting Limited
Priestley House
28-34 Albert Street
Birmingham
B4 7UD
UK
Tel : 21-616 1010
Fax : 21-616 1099

Gianfiero MEINERO
CGIL Savoma
17014 Cairo via Colla 19
Tel : 19-506260
Fax : 19-506261

Nicola MESSINA
FEDERCHIMICI
Via Academia, 33
10131 Milano
Italy
Tel : 2-268 10285
Fax : 2-268 10322

Michel MILLER
European Trade Union Confederation
Rue Montagne aux Herbes Potageres, 37
1000 Brussels
Belgium
Tel : 2-218 31 00
Fax : 2-218 35 66

Ugo MONTECCHI
CGIL Liguria
vico Tama 1
Genova
Italy

Raffaele MORESE
Confederazione Italiana Sindacati
Laboratori (CISL)
Via Po, 21
00198 Roma
Italy
Tel : 6-84 73 377
Fax : 6-84 73 203

Alessandro NOTARGIOVANNI
CGIL
Dip. to Ambiente
Corso d'Italia, 25
00198 Roma
Italia
Tel : 6 84 76 458
Fax : 6 884 56 83

Andrea OATES
Labour Research Department
78 Blakfriars Road
London
SE1 8HF
UK
Tel : 71-928 36 49
Fax : 71-928 06 21

Round Table on Industry, Social Dialogue and Sustainability

Susanne OBST
bei Ebrahimi
Bengelsdorfstr. 14
2000 Hamburg 71
Germany

Gina PARENTE-ADDISON
ECOTEC
avenue de Tervuren, 13B
1040 Brussels
Belgium
Tel : 2 732 7818
Fax : 2-732 7111

Jørn PEDERSON
European Foundation for the Improvement
of Living and Working Conditions
Loughlinstown House
Shankhill Co. Dublin
Ireland
Tel : 1-2826888
Fax : 1-2826456

Giovanna ROCCA ERCOLI
Direzione Generale Rapporti di Lavoro
Ministero del Lavoro e della Previdenza
Sociale
Via Flavia 6
00187 Roma
Italy
Tel : 6-46 83 1
Fax : 6-482 48 16

Giorgio RUSSOMANNO
Confortigianato
Settore Ambiente
Via San Giovanni in Laterano, 152
00184 Roma
Italy
Fax : 6-70452188

Franco SALVATORI
CGIL
Dip. to Ambiente
Corso d'Italia, 25
00198 Roma
Italia
Tel : 6 84 76 458
Fax : 6 884 56 83

Prof. Fernando SANTANA
Universidade Nova de Lisboa
Faculdade de Ciencias e Tecnologia
Departamento de Ciencias e Engenharia
do Ambiente
Quinta da Torre
2825 Monte de Caparica
Portugal
Tel : 29 54 464
Fax : 29 54 461

Antonio SANTOS ORTEGA
Universidad de Valencia
Aveniva Blasco Ibanez, 32
Valencia 46010
Spain
Tel : 96-386 44 14

Angelo SCARFINI
ASAP
via Due Macelli 66
00187 Roma
Italy
Tel : 6-673 4247
Fax : 6-673 4242

Round Table on Industry, Social Dialogue and Sustainability

Peter SCHLAFFER
Friedrich-Ebert-Foundation
Via del Babuino, 99
00187 Roma
Italy
Tel : 6 6788859
Fax : 6 6794897

Eberhard SCHMIDT
Herderstr. 48
2800 Bremen
Germany
Tel : 0421/77305

Massimo SETTIS
Unione Industriale
UFF. Energia e Ambiente
Via Fanti, 17
10128 Torino
Italy
Tel : 39-11 57 18 452
Fax : 39-11 57 18 217

Ana Cristina SEVERINO ALEIXO
UGT Portugal
Rua de Buenos Aires, 11
1200 Lisboa
Portugal
Tel : 397 65 03
Fax : 397 46 12

Margaret SWEENEY
SICA Innovation Consultants Ltd
44 Fitzwilliam Square
Dublin 2
Ireland
Tel : 613830
Fax : 613829

Arno TEUTSCH
Unione Italiana del Lavoro
Via Lucullo 6
00187 Roma
Italy
Tel : 6-47531
Fax : 6-4753280

Christina THEOHARI
Athens Labour Centre
Department of the Environment
48 B, 3rd September Str
10433 Athens
Greece
Tel : 1-88 41 818
Fax : 1-88 28 888

John P ULHØI
The Aarhus School of Business
Department of Organisation and Management
Ryhavevej 8
DK-8210 Aarhus V
Denmark
Tel : 45-15 86 55 88
Fax : 45-85 15 39 88

Efstathia VALIANTZA-AFTIA
FAINARETI
200, Kifissias Av
15231 Athens
Greece
Tel : (1) 8041820
Fax : (1) 8827002

Round Table on Industry, Social Dialogue and Sustainability

Eric VERBORGH
European Foundation for the Improvement
of Living and Working Conditions
Loughlinstown House
Shankhill Co. Dublin
Ireland
Tel : 1-2826888
Fax : 1-2826456

Silvano VERONESE
Unione Italiana del Lavoro
Via Lucullo 6
00187 Roma
Italy
Tel : 6-47531
Fax : 6-4753280

Juan VILA LOBOS
Ministere de l'Environnement
Cabinet Affaires Européennes
Rua Do Século 51
1200 Lisboa
Portugal
Tel : 351-1-3420831
Fax : 351-1-3472036

H E WILLIAMS
ECOTEC Research and Consulting Limited
Priestley House
28-34 Albert Street
Birmingham
B4 7UD
UK
Tel : 21-616-1010
Fax : 21-616 1099

Alwine WOISCHNIK
Union General de Trabajadores
c/Hortaleza, 88
28004 Madrid
Spain
Tel : 1-589 77 23
Fax : 1-589 76 03

ANNEX B

Round Table on Industry, Social Dialogue and Sustainability

ROUND TABLE ON INDUSTRY, SOCIAL DIALOGUE AND SUSTAINABILITY

BACKGROUND PAPER

1.0 THE OBJECTIVES AND INTERESTS OF THE EUROPEAN FOUNDATION

1.1 Background

The European Foundation for the Improvement of Living and Working Conditions, an autonomous EC institution, seeks to pursue and disseminate research specifically orientated towards improving living and working conditions in Europe. This work is administered around four-year programmes which are discussed with, and approved by, an Administration Board constituted by representatives of the European Commission, the Governments, employers' organisations and trade unions from the twelve Member States.

In its fourth work programme (1989-1992), which now comes to an end, six separate but interlinking areas of research and diffusion provide the framework for the Foundations' activity. A major theme running through these six areas of activity is the requirement to encourage and advance dialogue and communication between all parties with an interest in, and influence on, living and working conditions. This theme attains great significance in the context of a growing concern for the impact of industry on the environment, and the consequent need to improve the level of protection for the environment and for society.

The recognition of growing interest and action by both industry and workers in environmental matters was reflected in the organisation and conduct of a Round Table on the Role of Social Partners in Improving the Environment, held in Dublin in 1988. This Round Table highlighted the need for changes in attitude on the part of management and trade unions to the conduct of industrial activity in order to protect the environment, workers and the public. The Round Table concluded with:

• a call for a shift of emphasis from the promulgation of regulations and directives to achievement of adherence to their spirit as well as their letter;

• proposals for a redefinition of the nature of the environmental dialogue to embrace not only management and unions and industry generally, but also the European public;

• a consensus on the need to re-examine and revise the conceptual division between workers' and employers' environmental interests.

These conclusions are reflected in the fourth work programme (1989-1992), particularly in two of the six areas of European Foundation activity:

- developing industrial relations and participation (Area 1); and
- protecting the environment, the worker and the public (Area 4).

It is from work mainly undertaken within these two areas that the necessary evidence and experience, upon which to launch the Round Table in Rome, has been distilled.

1.2 Previous and Current Research

The research activities which have given rise to the findings and conclusions which are reflected in this paper fall into three broad areas.

1. The Education and Training of Personnel Concerned with Environmental Issues in Industry

This work provided a first opportunity to take a broad perspective as to the character of environmental pressures on industry and the consequent need for environmental training. The work examined not only the need for environmental training but also considered for the first time, the extent to which existing skills are adequate to cope with identifying and implementing the necessary changes in industrial activity to respond to environmental pressures. Complementary work to define the new competences which are required has been undertaken with CEDEFOP.

This broad area of work concluded that:

- inadequacies exist in the skills and competences of these categories of personnel;

- improvement in skills and competences are required to ensure that pressures for improved environmental performance by industry can be met;

- training provision and development of SMEs needs to recognise the difficulties facing SMEs in allocating time and resources to environmental training requirements;

- the development of policies to improve environmental management competences and industrial environmental performance needs to recognise that:

    * improved awareness and motivation among SMEs is a pre-requisite;
    * a wide range of organisations have a role to play in encouraging awareness and for improving environmental performance in industry, including the provision of training, information, advice and financial support;

* the encouragement of collaboration and co-operation between organisations is most effective at the local and regional levels;

* there is a particular need for these policies in Objective One regions, especially given the relatively weaker environmental infrastructure (waste treatment plants, collection systems etc) available to firms.

These conclusions form part of the background for the subsequent studies on the firm in its local environment and support a central tenet of the research programme that environmental protection and improvement are best achieved at the regional and local levels in a cooperation encompassing industry and its various organisations and associations, the social partners, public authorities and local communities.

2. The Firm in its Local Environment

As emphasised at the Round Table on the Role of the Social Partners in Improving the Environment, environmental issues and working conditions are often related and should, in such cases, be solved together. An integrated approach in this area is increasingly being applied in a few Member States. Such an approach recognises the pervasive character of environmental issues and as such should be encouraged widely in the Community in order to avoid unnecessary conflicts. Hence, it is essential that the relevant public authorities, companies and the social partners become more involved in a joint effort which will ensure that both environmental and working conditions' issues are being dealt with together, insofar as they are related.

This principle, combined with the conclusions from the first area of work, provides the basis for the second area of work designed to identify and review regional/local initiatives as the basis of a subsequent assessment of the need for improving existing initiatives and for creating new ones. The specific needs of SMEs, and of Objective One regions which may have particular difficulties in complying with environmental standards owing to lack of resources and expertise, are being examined explicitly by focusing on initiatives which address these particular issues.

The research has reached the stage of reviewing existing co-operation initiatives prior to an assessment of their effectiveness. Initial conclusions are, however, that the current regional and local attempts throughout the Community to improve the environmental performance of industry are desirable, and supportive of objectives to increase the awareness and competence of firms to address environmental issues.

3. Industrial Relations and the Environment

The third area of work which underpins the views expressed in this paper has focused on the extent to which the pervasive nature of environmental issues on industrial management require new forms of industrial relations. This work has demonstrated that there is a growing realisation amongst managers and workers that there is a common ground of industrial relations which seeks to encourage the prevention of harmful environmental impacts and to manage effectively the environmental risks of industrial activity. The identification of the extent of common ground differs between industries and between countries but there remains a growing realisation that since environmental pressures impact on all aspects of company activity that collective bargaining frameworks are right to expand to facilitate appropriate responses.

1.3 The Importance of Social Dialogue

The importance of encouraging and developing social dialogue in the context of greater environmental pressures has already been emphasised at the 1988 Round Table. Recent developments have added to the urgency of developing dialogue and models for resolving conflicts and establishing consensus. For example, increasing weight is being given by communities to the localised, often site specific, environmental impacts of industry: in turn requiring increasing efforts to resolve different interests. One response is to formalise the increasing use o environmental impact assessment. A further development is the increasing obligation on industry to publish information on corporate performance and environmental impact. The introduction of the Eco-Audit directive for example requires that those firms which volunteer to comply with the Directive publish records of their environmental performance. The increasing availability of, and access to, information because of its potential to highlight conflicts provides further impetus to the search for ways of mediating between different interests and establishing consensus.

1.4 The Objectives of the Round Table

The purpose of the Round Table is to provide an opportunity for Social Partners, government representatives and technical experts to consider what steps are necessary to promote the social dialogue and action, especially at a local level, which is so important to the improvement of industrial environmental performance and ultimately to the environmental sustainability of economic development. Moreover, given the relatively greater improvement in the environmental performance of industry required in Objective One regions, the Round Table provides a vehicle for considering the mechanisms which should be established to this end.

The importance of social dialogue, in summary, is to encourage a properly managed response to environmental pressures by contributing to:

- an awareness of environmental issues and their significance for industrial activity, amongst managers, workers, government and the public;

- a more informed response by industry to environmental pressures;

Round Table on Industry, Social Dialogue and Sustainability

- a more efficient environmental performance by industry by improving the external relationship with different interest groups (customers, suppliers, regulators) necessary to conduct business, and by improving internal industrial relationships necessary to secure the appropriate corporate response.

The Round Table is invited to focus, therefore, on:

- the role and influence of local co-operation initiatives as a vehicle for social dialogue;

- the need for additional initiatives and techniques for resolving conflicts and establishing consensus;

- the necessary management and procedural changes which are required by firms to respond effectively to environmental pressures;

- the necessary changes in industrial relations;

- the timescales which should be used when considering changes and the development of new policies;

- the issues and opportunities which need to be addressed in securing improvements of the environmental performance of industry in Objective One regions.

As an initial contribution to the Round Table we briefly consider, in Section 2.0, some of the key issues which underly the debate. Some observations on possibly policy directions are presented in the final section (3.0).

2.0 KEY ISSUES

2.1 Introduction

The Round Table is directed toward improving social dialogue which results in the improved environmental performance of industry and hence the long-term sustainability of economic development. On the basis of the work of the European Foundation to date there are a number of key issues which are fundamental to the discussions which should take place. These are discussed below.

2.2 Environmental Management : Managing Change

It is an underlying proposition of the Round Table and of this paper that environmental issues require fundamental changes in industrial management, in terms both of corporate policy and behaviour and in terms of specific production methods. The "environment" is not viewed as just another issue for companies to address and which can be tacked on to the bottom of the management agenda. It is viewed as representing a challenge of major proportions to the way in which industrial management and behaviour is conventionally conceived.

The character of environmental pressure which is impacting on industry is by now reasonably well understood in the abstract, resulting from a combination of regulatory and commercial pressures to improve environmental performance. The pressure on firms differs in different parts of the Community and for different types of firms. Key characteristics are:

- the growing awareness of larger firms that commercial success depends increasingly on better environmental performance;

- the growing impact of environmental regulations and subsequent enforcement. This is particularly important for SMEs;

- the increasing public concern for environmental improvements, tempered by the concern to protect employment and economic activity, evident throughout the Community;

- the difficulties facing firms, particularly in Objective One regions, in responding to environmental pressures caused by a lack of adequate environmental infrastructure.

Despite these different characteristics, and different models of corporate behaviour, the common message is that firms have to respond holistically. In practice this means that companies must:

- review from top to bottom how their activities impact, directly or indirectly, upon the environment;

- establish the necessary policies and procedures to adjust corporate behaviour to environmental pressures;

- ensure that the necessary skills and competences have been defined and are being met;

- monitor their environmental performance as a contribution to the required continuation of corporate environmental management.

We return later to innovations designed to assist companies in responding to these tasks, for example eco-auditing procedures.

2.3 Regulation Versus Self-Regulation

The policy context within which environmental performance is to be pursued is critical to determining the appropriate character of social dialogue. The main feature of the environmental policy context is the extent to which the enforcement of environmental regulation represents the driving force for change. Traditionally this approach has represented the main stream of environmental policy.

However, this is now being complemented by a growing interest and emphasis given to self regulation, encouraged by the development of instruments which use the market place to send the necessary signals to promote better environmental performance.

This change in emphasis recognises the enormous difficulties of enforcing environmental regulations across the population of companies in Europe, and is exemplified in the Fifth Environmental Action Programme, prepared by the European Commission. This programme emphasises the requirement for shared responsibility of public administrations, enterprise and the general public for actions to promote environmental quality. In particular, in relation to industry, the requirements for different organisations to "work together" and for a reinforcement of the dialogue with industry, are emphasised.

This policy context places an emphasis on anticipating and preventing environmental problems from arising in the first instance. Failure to do so will result in the risk of prosecution and closure and/or the loss of competitive advantage. The separation of regulatory functions from advisory functions, which have traditionally been combined, which is implied by the new approach means that enforcement activity is likely to be pursued vigorously. The presumption is that by placing the emphasis on self-regulation and commercial self interest, industry will respond in advance of the situation requiring enforcement action.

This emphasis on anticipation and prevention adds weight to the need for dialogue not only to increase the awareness of environmental pressures but also to improve the ability of firms to respond fully to the demand for change which environmental pressures bring about.

We highlight the need for improvement in the external relationships of the company, with customers, suppliers, regulators etc, and emphasise the role that these relationships can play in improving environmental performance. This is followed, in the next section, by a consideration of a parallel requirement to adopt internal relationships within the company, between managers and workers.

2.4 The Importance of Social Dialogue : External Communications

The firm, conventionally, has a set of relationships with external parties. These include: investors, customers, suppliers, regulators, planning authorities, the local community. These relationships provide a medium through which environmental pressures are communicated and amplified. For example, investors and customers increasingly demand higher environmental performance, with firms asked to comply with specified environmental management standards or risk losing finance and orders.

These external relationships, however, also offer a means by which firms can collectively take control of environmental issues and develop and plan appropriate responses. They provide an opportunity for firms to increase their awareness of environmental issues and to share experiences. Formal or semi-formal collections of firms are able to organise this process. These relationships can also provide the means of taking pro-active and preventative steps; by improving both the opportunity to highlight potential threats and to articulate the corporate response, for example by jointly commissioning advice and sponsoring investment in shared facilities.

The research of the European Foundation to date has shown up a wide variety of initiatives undertaken by firms and other parties which build upon external relationships. These initiatives contribute to the ability of firms, particularly SMEs, to improve their environmental performance. Two broad types exist:

- "Green" club; and
- Policy initiatives.

Green Clubs

We have, for convenience, termed those initiatives which aim to raise the general level of environmental awareness by means of an extensive networking and co-operation a "green club". These clubs tend to operate on a subscription basis with member firms receiving a number of benefits, which typically include participation in information exchange events, receipt of newsletters or memoranda, and access to information referral systems, usually the project officer.

The clubs operate on the basis that members, usually sponsoring companies, provide information and advice, either by responding to referral calls requesting assistance or by providing speakers to seminars etc or by contribution to newsletters or other written source material.

The extent of information exchange is usually viewed as being directly related to the scale of the co-operation; the more parties in the co-operation and the greater the number of events etc the greater the scale of information available and the more extensive the possibilities for diffusion.

The general purpose of these clubs and the possibility that firms make only small marginal changes in environmental management practices as a consequence of the co-operation makes it potentially difficult to assess how far this type of co-operation initiative is effective.

Policy Initiatives

The other group of initiatives which are distinguishable are those which focus on a more specific set of objectives related to a particular policy of environmental improvement; these may be either tri-party or multi-party co-operative initiatives. These initiatives have less of a requirement for extensive networks and a large number of partners. However, the initiatives covered in the Foundation enquiry strongly emphasise the value of the co-operation which does take place in support of the policy objectives.

Initiatives will typically comprise the public authorities, technical institutes and industry representatives (individual firms, trade associations, employer organisations). In the case where firms perceive the problem of inadequate environmental facilities, industry may be proactive in defining the initiative. In other initiatives, the direct involvement of firms is pursued to the point where a selection of individual firms, meeting specific criteria relating to the policy objectives, are identified and invited to join the co-operation. The value of the co-operation, as with the club, lies in encouraging firms to participate. In this context the co-operation is often viewed as an alternative or complementary approach to the strict enforcement of environmental legislation by pollution control inspectors, or to individual environmental investment requirements.

These initiatives, because they are more specific, have a greater requirement for technical advice and specialist assistance. The co-operation will typically include, either as a partner or by invitation, particular universities, research institutes or technical consultancies. Services to firms will be supplied freely, as part of the co-operation, or for a fee, sometimes paid for by the co-operation or by the public sector.

These initiatives, because they are regional/local in nature and often organised with a sectoral focus, are responsive to local conditions and circumstances and represent an important means of harnessing externally conducted social dialogue. These characteristics suggest that they have a particular role to play in promoting environmental performance in Objective One regions.

2.5 The Importance of Social Dialogue : Internal Communications

The ability of firms to respond to environmental pressures and to improve their environmental performance is ultimately determined by the managerial and technical capacity of the firm. Two closely related issues are instrumental in shaping this capacity.

2.5.1 Education and Training

The capacity of firms to manage change is determined by:

- an awareness of the pressures facing the firm;

- a motivation to secure higher levels of environmental performance (deriving from an assessment of regulatory and commercial pressures);

- an assessment of the required changes in corporate policy, and each aspect of corporate activity;

- the definition of required skills and competence;

- the acquisition, through training and/or consultancy, of the necessary skills.

The need for these determinants to be satisfied is not in question; the central issue is one of how best to satisfy these determinants. The research undertaken by the European Foundation suggests that the solution rests partly in developing external co-operation initiatives which are focused on one or more of these determinants. These initiatives have already been discussed. The central point is that re-defining skills to manage the response of firms represents a significant training and retraining effort.

2.5.2 Industrial Relations

The redefinition of skills on the basis of a corporate wide review of responsibilities has major implications for the conduct of industrial relations. In particular, the proactive development of corporate environmental policy, particularly where this arises because of public pressures, is likely to be of major interest to trade unions. Trade unions have been, and will continue to, demand an opportunity to influence this environmental policy.

This demand raises fundamental issues of control and responsibility; traditionally corporate policy and management activities are viewed as the prerogatives of senior managers. This dividing line is crossed by the development of corporate environmental policy.

The research of the European Foundation has highlighted the tensions which this revision in prevailing industrial relations can cause. Research also reveals the positive steps being taken to redefine the character of collective bargaining, and the responsibilities and interests of managers and workers.

2.6 Management Systems

The requirement to manage the process of change in response to environmental pressures gives rise to the need to introduce new management systems; with their associated goals. This need has been addressed, at least in the largest companies, by the introduction of environmental auditing processes and procedures, analogous to financial auditing, to assist environmental management.

The proposed EC-Directive on eco-auditing is one response to these pressures. In addition, environmental management standards are being developed by the standard institutions in a number of Member States. Furthermore, the eco-labelling proposals at the EC scale - and those which exist in certain Member States - also require that environmental management systems are in place.

There are a number of aspects to these developments. As firms learn to manage their processes more and more effectively and minimise waste and emissions, then the associated management systems become integrated with the quality, safety and operational systems of the company. However, in the context of eco-auditing, eco-labelling and meeting national standards there is an additional requirement for firms to have systems that are open and verified by independent assessors. These raise issues of commercial openness but also allow the elements of performance to enter into consumer, customer and investment considerations - which may be an important dimension of the effectiveness of market instruments. Management systems alone, of course, are just that. The degree to which they encourage achievement of higher standards and integrate regulatory requirements into the firm's operations vary from company to company. Finally, such systems can help to formalise the framework within which responsibilities for environmental performance are allocated and thus help to identify skill requirements and training needs.

3.0 POLICY DIRECTIONS

At the risk of imposing an early definition of the primary policy directions which the Round Table might pursue we briefly consider the likely future policy issues surrounding the debate.

- **Freedom of Information.** The increasing obligation on firms to publish information on their environmental performance increases the potential for explicating conflicts of interests between industry and government and between Social Partners. Consequently, models and techniques are required which establish consensus and facilitate mediation.

- **Timescales for Policy.** It is important to recognise that the debate, for much of industry, is still relatively new and that for many SMEs the debate is either alien or seen to be irrelevant. Moreover, defining and refining the necessary responses will take time. To the extent that social dialogue has initially to be based upon some recognition of the importance and magnitude of environmental pressures the development of both external and internal dialogues will take considerable time. The process of developing appropriate models for the character of response and character of the social dialogue must take time.

- **Economic and Environmental Policy Objectives.** Strongly related to the issues of timescale is the relative importance attached to short-term and long-term objectives. If it is accepted that the necessary responses by industry have short term costs, measured by reference to economic objectives, then the social dialogue must find ways of highlighting the long-term benefits not only in environmental terms but also in economic terms.

- **The Polluter Pays Principle.** The development of social dialogue and corporation will require resources. In some instances this can only come from public funds. This raises the issue as to whether the subsequent initiatives are deemed to be in contravention of the polluter pays principle. Whilst it is clear that any firm that can pay has no right to pollute freely the situation is more complex where the trade-offs between short-term private costs and long-term public benefits are ambiguous. The way in which the PPP is to be interpreted and applied in this context represents an important determinant of the character of social dialogue in the future. One direction is to widen the principle to include the idea of potential polluters paying for the prevention of pollution (the PPPPP!).

- **Economic Instruments.** The growing debate and increasing willingness to countenance economic instruments as a means of implementing environmental objectives represents a potent force for encouraging self-regulation and hence the demand for social dialogue. The debate as to how best to devise and introduce these instruments therefore overlaps with the debate concerning the promotion of social dialogue.

- **Objective One Regions.** The needs of Objective One regions to improve environmental performance is especially strong. The promotion of social dialogue has much to offer these regions. If this is the case then there must be a debate as to the best means by which social dialogue can be stimulated in these areas.

European Foundation for the Improvement of Living and Working Conditions

EUROPEAN ROUND TABLE ON INDUSTRY, SOCIAL DIALOGUE AND SUSTAINABILITY
A General Report

Luxembourg: Office for Official Publications of the European Communities

1993 — 85 pp. — 16.0 cm × 23.5 cm

ISBN 92-826-5817-1

Price (excluding VAT) in Luxembourg: ECU 10

Venta y suscripciones • Salg og abonnement • Verkauf und Abonnement • Πωλήσεις και συνδρομές
Sales and subscriptions • Vente et abonnements • Vendita e abbonamenti
Verkoop en abonnementen • Venda e assinaturas

BELGIQUE / BELGIË

**Moniteur belge /
Belgisch Staatsblad**
Rue de Louvain 42 / Leuvenseweg 42
B-1000 Bruxelles / B-1000 Brussel
Tél. (02) 512 00 26
Fax (02) 511 01 84

Autres distributeurs /
Overige verkooppunten

**Librairie européenne/
Europese boekhandel**
Rue de la Loi 244/Wetstraat 244
B-1040 Bruxelles / B-1040 Brussel
Tél. (02) 231 04 35
Fax (02) 735 08 60

Jean De Lannoy
Avenue du Roi 202 /Koningslaan 202
B-1060 Bruxelles / B-1060 Brussel
Tél. (02) 538 51 69
Télex 63220 UNBOOK B
Fax (02) 538 08 41

Document delivery:

Credoc
Rue de la Montagne 34 / Bergstraat 34
Bte 11 / Bus 11
B-1000 Bruxelles / B-1000 Brussel
Tél. (02) 511 69 41
Fax (02) 513 31 95

DANMARK

J. H. Schultz Information A/S
Herstedvang 10-12
DK-2620 Albertslund
Tlf. (45) 43 63 23 00
Fax (Sales) (45) 43 63 19 69
Fax (Management) (45) 43 63 19 49

DEUTSCHLAND

Bundesanzeiger Verlag
Breite Straße
Postfach 10 80 06
D-W-5000 Köln 1
Tel. (02 21) 20 29-0
Telex ANZEIGER BONN 8 882 595
Fax 2 02 92 78

GREECE/ΕΛΛΑΔΑ

G.C. Eleftheroudakis SA
International Bookstore
Nikis Street 4
GR-10563 Athens
Tel. (01) 322 63 23
Telex 219410 ELEF
Fax 323 98 21

ESPAÑA

Boletín Oficial del Estado
Trafalgar, 29
E-28071 Madrid
Tel. (91) 538 22 95
Fax (91) 538 23 49

Mundi-Prensa Libros, SA
Castelló, 37
E-28001 Madrid
Tel. (91) 431 33 99 (Libros)
     431 32 22 (Suscripciones)
     435 36 37 (Dirección)
Télex 49370-MPLI-E
Fax (91) 575 39 98

Sucursal:

Librería Internacional AEDOS
Consejo de Ciento, 391
E-08009 Barcelona
Tel. (93) 488 34 92
Fax (93) 487 76 59

**Llibreria de la Generalitat
de Catalunya**
Rambla dels Estudis, 118 (Palau Moja)
E-08002 Barcelona
Tel. (93) 302 68 35
     302 64 62
Fax (93) 302 12 99

FRANCE

**Journal officiel
Service des publications
des Communautés européennes**
26, rue Desaix
F-75727 Paris Cedex 15
Tél. (1) 40 58 75 00
Fax (1) 40 58 77 00

IRELAND

Government Supplies Agency
4-5 Harcourt Road
Dublin 2
Tel. (1) 61 31 11
Fax (1) 78 06 45

ITALIA

Licosa SpA
Via Duca di Calabria, 1/1
Casella postale 552
I-50125 Firenze
Tel. (055) 64 54 15
Fax 64 12 57
Telex 570466 LICOSA I

GRAND-DUCHÉ DE LUXEMBOURG

Messageries Paul Kraus
11, rue Christophe Plantin
L-2339 Luxembourg
Tél. 499 88 88
Télex 2515
Fax 499 88 84 44

NEDERLAND

SDU Overheidsinformatie
Externe Fondsen
Postbus 20014
2500 EA 's-Gravenhage
Tel. (070) 37 89 911
Fax (070) 34 75 778

PORTUGAL

Imprensa Nacional
Casa da Moeda, EP
Rua D. Francisco Manuel de Melo, 5
P-1092 Lisboa Codex
Tel. (01) 69 34 14

**Distribuidora de Livros
Bertrand, Ld.ª**
Grupo Bertrand, SA
Rua das Terras dos Vales, 4-A
Apartado 37
P-2700 Amadora Codex
Tel. (01) 49 59 050
Telex 15798 BERDIS
Fax 49 60 255

UNITED KINGDOM

HMSO Books (Agency section)
HMSO Publications Centre
51 Nine Elms Lane
London SW8 5DR
Tel. (071) 873 9090
Fax 873 8463
Telex 29 71 138

ÖSTERREICH

**Manz'sche Verlags-
und Universitätsbuchhandlung**
Kohlmarkt 16
A-1014 Wien
Tel. (0222) 531 61-0
Telex 112 500 BOX A
Fax (0222) 531 61-39

SUOMI

Akateeminen Kirjakauppa
Keskuskatu 1
PO Box 128
SF-00101 Helsinki
Tel. (0) 121 41
Fax (0) 121 44 41

NORGE

Narvesen information center
Bertrand Narvesens vei 2
PO Box 6125 Etterstad
N-0602 Oslo 6
Tel. (2) 57 33 00
Telex 79668 NIC N
Fax (2) 68 19 01

SVERIGE

BTJ
Tryck Traktorwägen 13
S-222 60 Lund
Tel. (046) 18 00 00
Fax (046) 18 01 25

SCHWEIZ / SUISSE / SVIZZERA

OSEC
Stampfenbachstraße 85
CH-8035 Zürich
Tel. (01) 365 54 49
Fax (01) 365 54 11

CESKOSLOVENSKO

NIS
Havelkova 22
13000 Praha 3
Tel. (02) 235 84 46
Fax 42-2-264775

MAGYARORSZÁG

Euro-Info-Service
Pf. 1271
H-1464 Budapest
Tel./Fax (1) 111 60 61/111 62 16

POLSKA

Business Foundation
ul. Krucza 38/42
00-512 Warszawa
Tel. (22) 21 99 93, 628-28-82
International Fax&Phone
(0-39) 12-00-77

ROUMANIE

Euromedia
65, Strada Dionisie Lupu
70184 Bucuresti
Tel./Fax 0 12 96 46

BULGARIE

D.J.B.
59, bd Vitocha
1000 Sofia
Tel./Fax 2 810158

RUSSIA

**CCEC (Centre for Cooperation with
the European Communities)**
9, Prospekt 60-let Oktyabria
117312 Moscow
Tel. 095 135 52 87
Fax 095 420 21 44

CYPRUS

**Cyprus Chamber of Commerce and
Industry**
Chamber Building
38 Grivas Dhigenis Ave
3 Deligiorgis Street
PO Box 1455
Nicosia
Tel. (2) 449500/462312
Fax (2) 458630

TÜRKIYE

**Pres Gazete Kitap Dergi
Pazarlama Dağitim Ticaret ve sanayi
AŞ**
Narlibahçe Sokak N. 15
Istanbul-Cağaloğlu
Tel. (1) 520 92 96 - 528 55 66
Fax 520 64 57
Telex 23822 DSVO-TR

ISRAEL

ROY International
PO Box 13056
41 Mishmar Hayarden Street
Tel Aviv 61130
Tel. 3 496 108
Fax 3 544 60 39

CANADA

Renouf Publishing Co. Ltd
Mail orders — Head Office:
1294 Algoma Road
Ottawa, Ontario K1B 3W8
Tel. (613) 741 43 33
Fax (613) 741 54 39
Telex 0534783

Ottawa Store:
61 Sparks Street
Tel. (613) 238 89 85

Toronto Store:
211 Yonge Street
Tel. (416) 363 31 71

UNITED STATES OF AMERICA

UNIPUB
4611-F Assembly Drive
Lanham, MD 20706-4391
Tel. Toll Free (800) 274 4888
Fax (301) 459 0056

AUSTRALIA

Hunter Publications
58A Gipps Street
Collingwood
Victoria 3066
Tel. (3) 417 5361
Fax (3) 419 7154

JAPAN

Kinokuniya Company Ltd
17-7 Shinjuku 3-Chome
Shinjuku-ku
Tokyo 160-91
Tel. (03) 3439-0121

Journal Department
PO Box 55 Chitose
Tokyo 156
Tel. (03) 3439-0124

SINGAPORE

Legal Library Services Ltd
STK Agency
Robinson Road
PO Box 1817
Singapore 9036

AUTRES PAYS
OTHER COUNTRIES
ANDERE LÄNDER

**Office des publications officielles
des Communautés européennes**
2, rue Mercier
L-2985 Luxembourg
Tél. 499 28 1
Télex PUBOF LU 1324 b
Fax 48 85 73/48 68 17

10/92